Mawulolo Koevi
Iléri Dandonougbo

Cycling in the Autonomous District of Greater Lomé (Togo)

Mawulolo Koevi
Iléri Dandonougbo

Cycling in the Autonomous District of Greater Lomé (Togo)

A little-used mode of transport

ScienciaScripts

Imprint

Any brand names and product names mentioned in this book are subject to trademark, brand or patent protection and are trademarks or registered trademarks of their respective holders. The use of brand names, product names, common names, trade names, product descriptions etc. even without a particular marking in this work is in no way to be construed to mean that such names may be regarded as unrestricted in respect of trademark and brand protection legislation and could thus be used by anyone.

Cover image: www.ingimage.com

This book is a translation from the original published under ISBN 978-620-6-72104-8.

Publisher:
Sciencia Scripts
is a trademark of
Dodo Books Indian Ocean Ltd. and OmniScriptum S.R.L publishing group

120 High Road, East Finchley, London, N2 9ED, United Kingdom
Str. Armeneasca 28/1, office 1, Chisinau MD-2012, Republic of Moldova, Europe
Printed at: see last page
ISBN: 978-620-8-07580-4

BICYCLES IN THE AUTONOMOUS DISTRICT OF GRAND LOME (TOGO): A MODE OF TRANSPORT WITH LOW LEVELS OF USE

Summary

Since 2000, cities in sub-Saharan Africa have entered the arena of metropolization. The sprawl of these cities and the concentration of people within them are creating major mobility challenges. To solve these problems, public authorities have embarked on major works policies, notably the redevelopment and construction of road infrastructure. From 2010 to 2018, 810 km of roads will be upgraded in the Greater Lomé Autonomous District. In this metropolis, cyclists are poorly represented and sidelined during the construction of these road infrastructures.

The aim of this study is to analyze the reasons for the low use of bicycles in the Autonomous District of Greater Lomé, in order to highlight strategies for improving this soft mode of transport in the study area. The methodology used in this research is based, on the one hand, on the literature to examine the various theories governing the analysis of issues relating to cycling mobility, and on the other hand, on the analysis of field realities through direct observations, surveys and interviews with various stakeholders. Overwhelming heat, longer distances between city centers and suburbs, negative perceptions of cycling and the absence of cycle paths are the main factors behind the low use of bicycles in Greater Lomé. In this conurbation, 99.37% of streets have no bicycle lanes. Those that do exist lack pavement markings and are impassable, as they are occupied by small businesses.

Key words: District Autonome du Grand Lomé (DAGL), soft mobility, urban transport, cyclists, bicycle.

Table of contents

Introduction

In sub-Saharan Africa, mobility needs are a major issue for populations and municipalities alike. The need for motorized traffic is a central concern. The increase in road traffic is exacerbating mobility problems. Motorized transport has a negative impact on the environment. In 2018, according to the United Nations Environment Programme (UNEP)[1], in developing countries, 90% of air pollution in urban areas is attributable to vehicle emissions. In Africa, 176,000 deaths are caused by outdoor air pollution every year (WHO, 2018, p. 16). This situation is receiving a lot of attention from international organizations such as the United Nations, the World Bank, the IMF, the WHO and is explained by high motorization. In Togo, according to the national greenhouse gas inventory report drawn up by the Ministry of the Environment and Forest Resources (MERF), emissions from the energy sector for 2018 were 2007.05 Gg [2][3]*of CO2, with road transport alone emitting 1471.58 Gg, or 56% of this gas. In the 1970s-1980s, "motorcycles were only used by civil servants, school principals, educational advisors and a few prefectural officials" (D. K. Suka, 2021, p. 135). But in the early 2000s, with the democratization of the motorcycle and the absence of a public transport service, motorcycles of Chinese and Indian origin such as Sanya, Sanili,

---

[1] UNEP: The Emissions Gap report 2018, Nairobi, United Nations Environment Programme,

[2] Gigagram: system mass measurement unit international worth 10^9 grams or 10^6 kilograms

Lifan, Nanfang, Jincheng, Léopard, Apsonic, TVS and Haojue invaded the streets of the Greater Lomé Autonomous District (D. K. Suka, 2021, p. 206).

In the Greater Lomé District, the growth of motorized two-wheelers is increasing mobility-related problems. This dominance of motorized two-wheelers is redefining the geography of urban transport in this area (A. Guezerre, 2013, p. 42). In 2015, according to the diagnosis of the Schéma Directeur d'Aménagement et d'Urbanisme de Lomé, 10% of heads of household had a bicycle for their daily commute. The modal share of motorized two-wheelers was 65%, 24% for cars and 1% for trucks. Inhabitants of this urban center show little interest in cycling. As is the case in large sub-Saharan conurbations, this means of travel is the "great absentee" from the streets of Greater Lomé. Road counts carried out by A. Guezere (2008, p. 234) show that the modal share of bicycles is around 5%. In 2021, this mode of mobility was used by just 1.1% of Greater Lomé's inhabitants (D. K. Suka, 2021, p. 134). Bicycle use is therefore low. Given this situation, the question is: what factors underlie the low use of bicycles in Greater Lomé?

To answer this question, it is necessary to analyze the respective effects of the natural environment, road conditions, high motorization and people's perception of cycling mobility in the Autonomous District of Greater Lomé. This study has two main focuses. It presents the material and methods, and the factors explaining the low level of bicycle use in the Autonomous District of Greater Lomé.

1. Materials and methods

1.1. Study framework

Greater Lomé lies between 6°10' and 6°25' north latitude and 1°05' and 1°25' east longitude in the maritime region, one of Togo's five economic regions (Map 1).

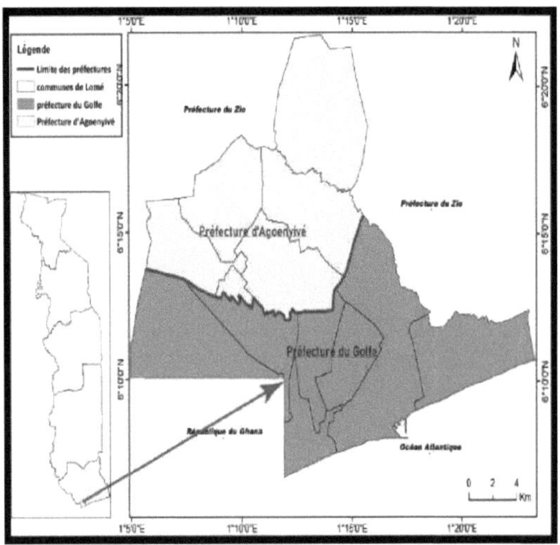

Map 1: Location of Greater Lomé

Source: M. Koevi, 2023 According to Map 1, the Lomé conurbation includes the prefectures of Golfe and Agoenyivé. Located on the Atlantic Ocean, in the Gulf of Guinea, Greater Lomé is bordered to the north by the Zio prefecture, to the east by the Lakes prefecture, and to the west by Ghana. It is the country's largest urban center in terms of population, surface area and concentration of socio-economic activities, home to 2.2 million people, or 25% of the total population (RGPH-5, 2022). The Greater Lomé Autonomous District covers an area of 42,560 ha and comprises 13 communes. It is experiencing strong demographic and spatial growth, lengthening center-periphery distances and stimulating the use of motorized modes of transport.

1.2. Methodological framework

To answer the question of this study and achieve the main objective, a methodological approach is necessary. It includes documentary research, direct observations, interviews and questionnaire surveys. The information gathered concerns mobility in Lomé, in particular cycling mobility. Documentary research is complemented by participant observations, enabling us to assess the state of the road network and the situation of cyclists in this metropolis.

The field surveys were conducted from February 5 to 27, 2023, a total of 23 days. The cyclists interviewed were those on boulevard Général Gnassingbé Eyadema, boulevard du 13 janvier, boulevard Jean Paul II, and the route du grand contournement de Lomé. This choice enables us to analyze the difficulties of cycling mobility in Lomé and to understand the traffic of non-motorized two-wheelers. To obtain a representative sample, 150 cyclists were randomly interviewed on these different roads. Cyclists from various neighborhoods, such as Baguida and Adétikopé, were included in the sample (Table 1).

Table 1: Distribution of respondents

Main roads	*Number of respondents*
Downtown - Adétikopé axis (Boulevard Gnassingbé Eyadéma),	35
Downtown - Sanguéra axis	27
Downtown -Baguida axis	29
Centre-Ville - Kégué axis (Boulevard Jean Paul II)	28
Lomé ring road	31
Total	150

Source: Field work, 2023

According to Table 1, five (5) roads were selected for the surveys. In order to better analyze urban mobility conditions and understand the obstacles to cycling mobility in Greater Lomé, participant observation was carried out by shuttling along the area's main arteries on foot and by bicycle. Three main routes were selected:

- ➢ the first axis is Agoè Assiyéyé-carrefour 2 Iions-Carrefour Caméléon GTA-Voie de la Nouvelle Présidence-Kégué-Hedzranawoe (Marché Hedzranawoe);

- ➢ the second line concerns: GTA- Boulevard Eyadéma - Université de Lomé-Lycée de Tokoin- Colombe de la paix -

Avenue Maman N'danida - Deckon-Grand Marché ;

➢ the third artery is the coastal route (RN2) from Kodjoviakopé to Baguida.

This exercise analyzed temporality, travel attitudes and cohabitation between cyclists and other users (pedestrians, motorcyclists, tricycle drivers, cars, buses, trucks). Data were processed using QGIS 2.14, SPSS, Excel and Word. QGIS was used to produce maps, SPSS to calculate averages, and Excel and Word for graphs, tables and text. The digital camera was used to illustrate the various situations showing the difficulties of cycling mobility in Greater Lomé through photos. The recorder was used for the interview sessions. A flow-counting operation was carried out in February 2023, a period outside school vacations. This methodology yielded the following results.

2. Results

The results of the survey address the factors behind the low use of bicycles in the Greater Lomé Autonomous District.

2.1. Cycling: the least-used travel mode in the DAGL

In metropolitan Lomé, bicycle use is very limited. It remains the mode of transport least used in this urban center (figure 1).

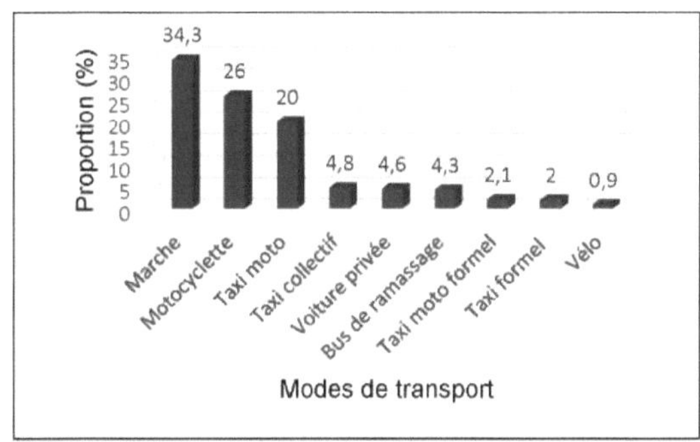

Figure 1: Share of each mobility mode in Greater Lomé

Source: Field work, 2023

Analysis of the results presented in figure 1 shows that the means of transport least used to get to work are the formal motorcycle cab (2.1%), the formal cab (2%), and the bicycle (0.9%). Bicycles are the least used mode of transport in Greater Lomé. This lack of interest in non-motorized two-wheelers among Greater Lomé city dwellers is due to the combined effects of several factors.

2.2. Factors behind the low use of bicycles in Greater Lomé

Unlike rural areas, where people prefer to get around by bike, this means of transport is less popular in sub-Saharan Africa's major cities. Despite its preponderance in these urban centers from the 1920s to the late 1960s, the bicycle's share of urban travel has been declining. In 1928, there were a total of 1,772 two-wheeled means of transport in Togo, including 731 bicycles (97.69%) and 41 motorcycles (2.31%). In Lomé, as in all Togo's cities, bicycles dominated the transport sector. In Greater Lomé, there were 637 bicycles (61.3%), 15 motorcycles (1.44%), 96 passenger cars (9.2%), 14 vans (1.35%), 77 500-1,000 kg trucks (7.4%), 194 trucks (18.7%) and 7 ANT tractors (0.6%).

(K. Kouzan, 2022, p. 31). The reasons for rejecting this mode of travel are diverse (Figure 2).

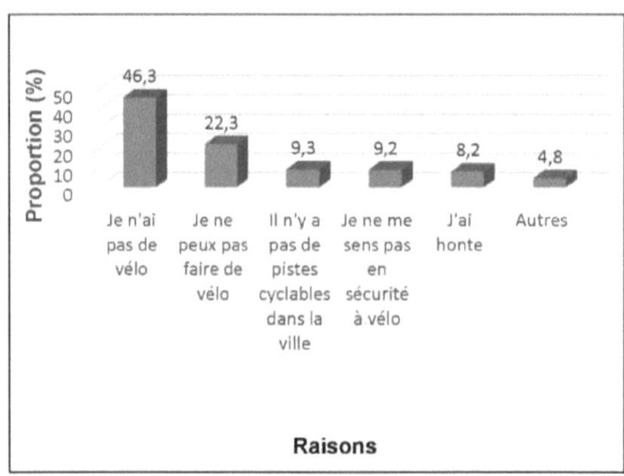

Figure 2: Reasons for rejecting bicycles in Greater Lomé

According to figure 2, there are several reasons why city dwellers in Greater Lomé abandon bicycles. These reasons have more to do with bike ownership and control. For 46.3% of respondents, the reason for not using a bicycle is that they don't have one. This situation does not depend on low purchasing power, but on a lack of desire to have one, because in this city, the price of a bicycle is within the population's reach (sDAU, 2016, p. 75). The price varies between 25,000 F CFA and 80,000 F CFA. The cost of a bicycle is twelve (12) times lower than that of a motorcycle, which varies between 400,000 FCFA and 2,500,000 F CFA. However, Loméans prefer motorized two-wheelers. In a context marked by economic precariousness, it is paradoxical that the constraints of bicycle use are less a question of economic possibilities than of desire. This situation confirms that purchasing power is not the main factor behind the low level of bicycle use in this metropolis.

A number of factors are behind Greater Lomé residents' lack of interest in this mode of mobility. They are natural, politico-strategic and socio-economic.

2.2.1 A mode of transport limited by the physical environment

The physical features of the Autonomous District of Greater Lomé have a major impact on its layout and transport conditions. Overwhelming heat, sandy and clayey soils, and double-peak rainfall hamper cyclists' mobility. Temperatures range from 25°C to 31°C, reaching a maximum of 35°C in 2022 (DMN). These high temperatures make cycling difficult, as cyclists avoid sweating by choosing motorized modes of transport. In addition, average precipitation is 874.8 mm/year (DMN, 2020).

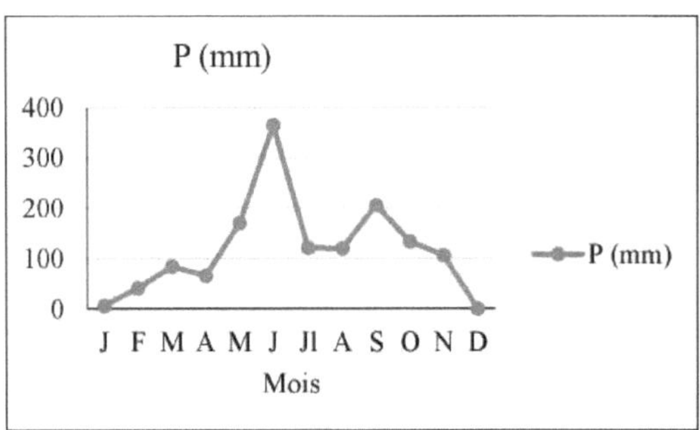

Figure 3: Annual rainfall trends in Lomé (2022)

Source: DMN, 2022

Rainfall in the Greater Lomé Autonomous District is bimodal. These rains cause flooding in the poorly permeable soils, complicating bicycle travel. Cyclists generally avoid going out during the rainy season (Plate 1).

Plate 1: Flooded streets in the Zongo district of Agoenyivé

Source: Field work, June 2023

The photos in Plate 1 show flooded streets in Zongo, in the commune of Agoenyivé 4, in the north of the Autonomous District of Greater Lomé. The poor condition of the streets makes mobility difficult, especially for motorists and cyclists. After rain, water stagnates for between 8 and 24 hours, or even longer, depending on the topography. In low-lying areas, stagnation lasts more than 24 hours. In the rainy season, puddles hide potholes and gullies, making streets impassable for bicycles, especially on unpaved roads. Water and mud cause cyclists to slip and get dirty. Thunderstorms with strong winds reduce visibility, hampering cyclists and sometimes causing falls. These conditions hamper cycling mobility in Greater Lomé. In addition to physical obstacles, high levels of motorization also limit cycling.

2.2.2. *High motorization of Greater Lomé: a limiting factor for cycling*

One of the reasons for the low use of bicycles in the Autonomous District of Greater Lomé is the high level of motorization. Between 2000 and 2020, the average radius of this city increased from 10 km to 25 km (D. K. Suka, 2021, p.167). This increase in distances has led to innovations in transport. The number of cars and motorcycles in this conurbation is growing considerably, especially with the boom in individual motorization (A. Guezere 2012, p. 45). Between 2000 and 2020, the number of motorized two-wheelers rose sharply (figure 3).

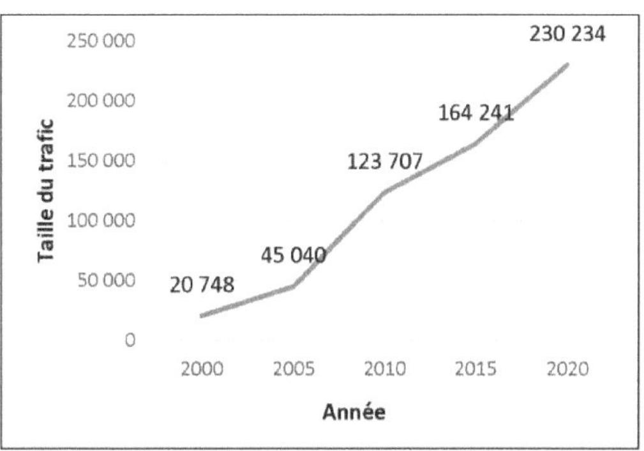

Figure 4: Number of motorized two-wheelers in the Greater Lomé Autonomous District

Source: BOAD-SOTED (2015), DTRF (2020)

Figure 3 shows the evolution of motorized two-wheelers in the Greater Lomé Autonomous District from 2000 to 2020. The year 2005 marks the start of a vertiginous growth in motorized two-wheelers. From 2000 to 2005, the number of two-wheelers more than doubled. It went from 20,748 motorcycles to 45,040. Between 2005 and 2010, growth was even more meteoric. The fleet reached 123,707 motorcycles in 2010 and 164,241 in 2015. By 2020, there will be 230,241 motorcycles in Greater Lomé, an increase of 184,241 in 20 years. Apart from two-wheelers, cars account for a significant share of motorized transport in this area (figure 4).

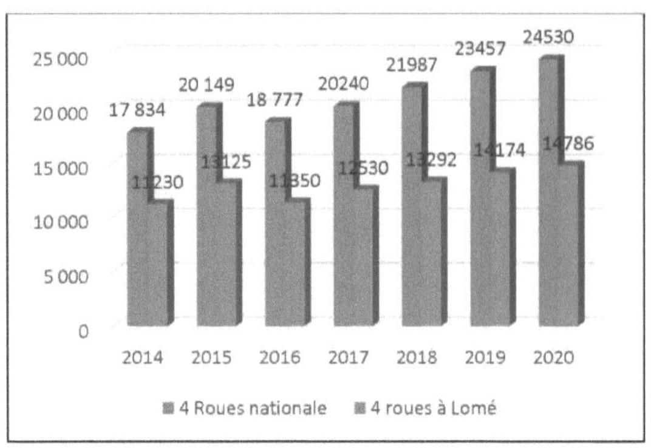

Figure 5: Growing car fleet

Source: DTRF database, 2020

Figure 4 shows the continued growth of motorized vehicles in Greater Lomé from 2010 to 2020, with more than half of the national car fleet concentrated in this area. This increase is due to urban sprawl and the desire of residents to own their own vehicle. However, this motorization has not been accompanied by improvements to road infrastructure or the construction of bicycle paths, exposing cyclists to the risk of accidents. In 2021, 7,130 accidents involved 12,597 vehicles, including 7,442 two-wheelers, causing 576 deaths and 9,514 injuries. According to surveys, 37% of cyclists have been hit by a motorized vehicle at least once. Poor road conditions and the absence of cycle paths explain the low use of bicycles in Greater Lomé.

2.2.3. *The poor state of the roads and lack of space for bicycles*

Despite policies to develop and maintain road infrastructure, the road network in the Autonomous District of Greater Lomé is inadequate and deficient. The new roads are narrow, with widths ranging from 12 to 16 meters. Cyclists are marginalized on these narrow roads, as there is no dedicated space for them. The shortcomings of these roads can also be seen in their road surfaces. In this urban center, only 14.35% of streets are paved, of which 39.89% are in good condition. These shortcomings make it difficult for cyclists to get around, and diminish the interest of Greater Lomé residents in cycling. Transport in general in the Autonomous District of Greater Lomé is marked by poor road layout. All roads are unpaved and in a dilapidated state (Table 2).

Table 2: Street conditions in the Autonomous District of Greater Lomé

Coating type	Status			Total (km)
	Good ( )^{km}	Average (km)	Bad ( )^{km}	
Bitumen	39,064	62,637	10,266	111,967
Keypad	15,826	9,557	2,35	25, 618
Laterite refill	1,99	5,476	3,583	9, 259
Earth	2, 521	77,115	731,984	811, 620
Total	57,611	154,78	746,068	958, 466

Source: AG7 Report, DGIEU, 2020

Table 2 shows that only a third of paved streets are in good condition, while 90.18% of dirt streets are in poor condition, representing 811.620 km, or 84.67% of the total road network. The data in Table 1 show that of a total road length of 958.466 km, including 111.967 km of asphalt roads, 746.068 km of the structural network is in poor condition, with only 57.611 km in good condition in the Greater Lomé Autonomous District. Drainage systems are in poor working order, and pavement protection is inadequate, notably due to the lack of shoulders. According to field surveys, 71% of cyclists often use paved streets, but only 34% feel comfortable on them, which puts cycling at a disadvantage. What's more, the absence of cycle paths is a real obstacle to cyclists' mobility, forcing them to share the road with motorized users.

Road infrastructure in the Autonomous District of Greater Lomé is unevenly distributed. Road density varies from one commune to another (Map 2).

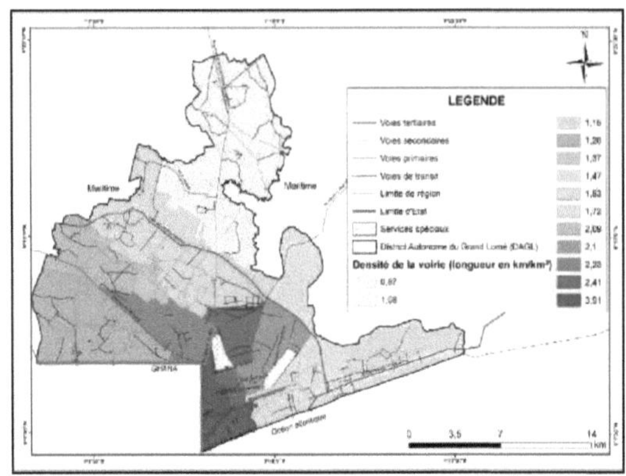

Map 2: Road density in the DAGL

Source: M. Koevi, 2023

Map 2 shows road density in the Autonomous District of Greater Lomé. It shows an uneven distribution of road density across the study area. The lower town areas (near the Atlantic coast) show a high road density (over 2 km per km2). This zone includes the city center and surrounding area, where economic activity is concentrated and mobility is crucial to daily operations, especially during rush hours.

The shortcomings of these infrastructures are compounded by poor street lighting, which hinders cycling at night.

2.2.4. *Poorly lit streets: a limiting factor for cyclists*

Another mobility problem in Greater Lomé is street electrification. The capital's streets lack lighting, as shown in Plate 2.

Plate 2: Poor street lighting in the Autonomous District of Greater Lomé

Source: Field work, June 2023

The photos in Plate 2 show poorly lit or unlit streets, a problem for cyclists at night. Since 2008, 40% of the streets in Greater Lomé have been lit, but the streetlights are not always working regularly. The city center is better lit than outlying areas such as Aflao Sagbado and Baguida, where only 29,490 linear meters of streets are lit. In these areas, 66.7% of cyclists use roads without electricity poles. What's more, 63% of bicycles have no headlights, limiting night-time mobility. Inadequate lighting explains the low use of bicycles at night, exacerbated by urban sprawl and longer distances.

2.2.5. Cycling in the Autonomous District of Greater Lomé: a local means of transport

The pace of Lomé's spatial growth has given it the character of a spread-out city, where the distance between center and periphery is continually increasing. Dormitory districts are moving further away from business districts.

The actual radius of Greater Lomé varies between 15 km and 25 km, while cycling in an urban area is effective for distances of less than 10 km. Figure 5 shows the average distances covered by bicycle users in the Greater Lomé Autonomous District to get to their places of work.

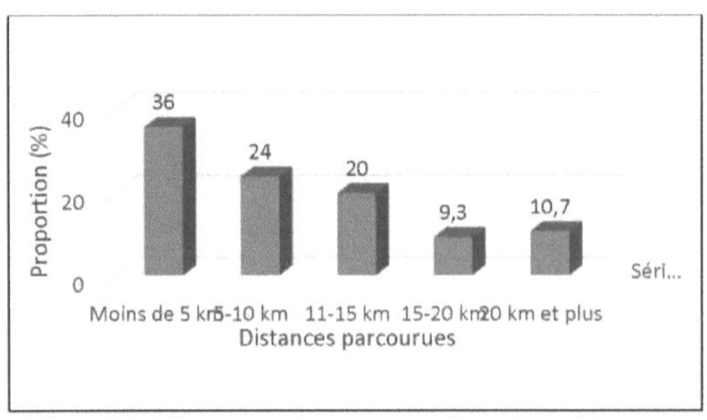

Figure 6: Breakdown of cyclists surveyed by distance travelled from home to work or school

Source: Field work, 2023

According to Figure 5, 60% of cyclists

travel less than 10 km a day to get to work or school. The number of cyclists decreases with increasing distance, with only 10.7% travelling 20 km or more daily. This shows the impact of urban sprawl on the low use of bicycles in Greater Lomé. However, these factors alone do not explain this low level of use; it also depends on the perception of local residents.

2.2.6.Loméans' perception of bicycles: a degraded image

In large sub-Saharan cities, the bicycle is seen as a means of transport for the poor, for city dwellers who can't afford a motorcycle or a car. In the Autonomous District of Greater Lomé, people perceive the bicycle as an obsolete means of transport and a sign of "poverty and rurality" (figure 6).

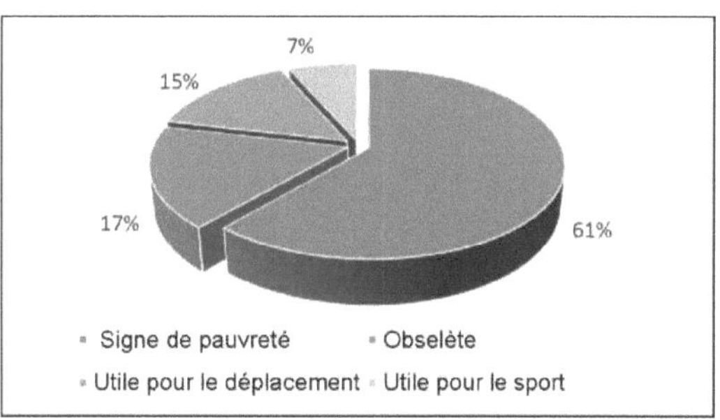

Figure 7: Loméans' perception of cycling

Source: Field work, 2023

Figure 6 shows that the people of Lomé associate bicycles with poverty, with 61% of those surveyed believing that only the poor use them. Only 15% of cyclists see it as a useful means of transport and 7% as a means of sport. Owning a motorcycle or a car is seen as a sign of success. Among cyclists, 81.3% want to change their mode of transport, and 22.7% want to escape the scorn of other users. What's more, 74.7% have been the victims of derogatory remarks, often called "poor", "villagers" and "less civilized".

3. Discussion

The low use of bicycles in the Autonomous District of Greater Lomé is due to natural, economic, strategic and social factors. Socially disadvantaged, the bicycle remains negatively perceived, which limits its adoption as a mode of soft mobility in sub-Saharan metropolises. C. Aholou and K. H. Logan (2021, p. 8) point out that cycling is often associated with poverty and contempt in Lomé traffic. This perception is comparable to that observed by M. Cusset (1995, p. 88) in West Africa, where riding a bicycle is considered degrading, in contrast to certain parts of Asia where cycling is more highly valued than walking. D. Olvera and D. Plat (1996, p. 298) note that cycling is particularly appreciated in rural contexts. P. Pochet (2002, p. 6) explains the absence of bicycles in sub-Saharan capitals by their perception as "symbols of rurality and poverty".

Social perception is not the only factor explaining the low use of bicycles. Insecurity is also a major factor. This insecurity is due in part to the relentless growth in motorized traffic. The mixing of motorized and vehicular traffic puts cyclists at a dead end on the road, with a constant risk of accident. The writings of A. Guézéré (2012, p. 70) on motorized travel in Lomé go in the same direction when he states that while relations are more or less acceptable between motorcycle and car drivers, cyclists and pedestrians are seen as sources of inconvenience.

A. Passoli et al (2024, p. 21) explain that constraints such as the lack of crossings on main roads, motorized drivers' failure to comply with the Highway Code, and their poor behavior have a negative impact on the mobility of active mode users. Their study reveals that in Greater Lomé, 60% of bicycle trips cover distances of 10 km or less. The use of bicycles is sought after as a fast, inexpensive and convenient mode of travel for short and medium-distance journeys (SDAU, 2014, p. 217). This is not shared by C. Aholou, K. H. Logan (2021, p. 8), who instead explain that:

Cyclists' journeys in Lomé are not often local. The majority of trips go beyond the cyclist's home district, with only % saying they travel less than 5 km from home to work. The average journey concerns % of those surveyed and is between 5 and 10 km.

Conclusion

This study of cycling mobility in the Autonomous District of Greater Lomé has enabled us to identify the factors that explain the low use of bicycles. It emerged that the low use of this means of transport is due to the constraints of the natural environment, such as oppressive heat and rain, the state of the roads, high levels of motorization, urban sprawl followed by longer distances, and Loméans' negative perception of cycling mobility. The image of cycling in the Autonomous District of Greater Lomé has deteriorated. It is used by people living in peri-urban areas, who are less affluent and less urban than those living in the city's central districts. According to field surveys, the main users of bicycles in this metropolis are schoolchildren (35%), students (21%) and apprentices (16%). Cycling is reserved for local trips, as 60% of cyclists surveyed cover a distance of 10 km or less by bike for their daily commute. At a time when sustainable development is at the forefront of the debate, this situation is of particular concern to the State and public authorities. Promoting cycling mobility becomes the first option for reducing the environmental pollution for which road transport is responsible through the direct emission of 10.5% of global CO_2, and 74.6% of these emissions are transport-related, according to the *World Resource Institute* (WRI, 2016). With this in mind, the aim is to encourage the use of bicycles by prioritizing the development of infrastructures adapted to the safe use of bicycles in this urban space, by taking measures to facilitate their acquisition and by reducing motorized travel. The challenge is to alleviate the mobility problems that plague this metropolis in environmental, economic and social terms. Promoting and developing cycling mobility in the Autonomous District of Greater Lomé will contribute to the sustainability of mobility and make cycling a particularly effective complementary mode of transport in the peri-urban areas of this urban center.

Bibliographical reference

AHOLOU Coffi, LOGAN Koffi Hubert, 2021, "Usage du vélo dans la mobilité active à Lomé : au-delà des contraintes, le bénéfice santé", In *Revue Espace,*

Territoire, Sociétés et Santé, pp.7-20 CUSSET Jean-Michel et al., 1995, "Les Transports urbains non motorisés en Afrique sub Saharienne. Le cas du Burkina Faso", coll. *SITRASS,* 138 p.

DIAZ OLVERA Lourdes, PLAT Didier, 1994, "Usages et images du vélo à Ouagadougou",

Research Transport Safety, n° 45

pp. 45-54.

FAGBEDJI Kodjo Gnimavor, HETCHELI Follygan, DANDONOUGBO Iléri, 2021,

"L'éclairage public des rues dans les espaces périurbains de Lomé (Togo) : de l'inégalité spatiale à l'injustice", in AKAKPO Yaovi (dir.), *Aménagement du territoire et sentiers d'économie en Afrique : fonction de bricolage.*

technologique, innovations sociales en Afrique, Collection Études Africaines, l'Harmattan, Condé-en-Normandie

(France), March 2021, ISBN 978-2-34322224-0, pp. 49-74.

GUEZERE Assogba, 2009,

"Complémentarité et intégration spatiale des transports artisanaux à Lomé" In Revue de Géographie Tropicale et d'Environnement, n° 1, EDUCI, pp. 51-63. KOUZAN Komlan, 2022, "L'utilisation de la bicyclette et de la motocyclette au Togo à l'époque coloniale (1884-1960)", In *Géotransports*, n° 17-18, pp. 27-40.

LAROSE Frédéric, 2011, "La pertinence du vélo en ville, Le vélo au cœur des politiques de mobilité durable", http://base.citego.info/fr/corpus analyse/fi che-analyse-65.html.

LOGAN Koffi Hubert, 2020, *Modes doux de mobilité urbaine: perception et contraintes des déplacements à vélo dans le grand Lomé,* Master's thesis, Geography Department, University of Lomé, Lomé, 106 p.

WHO, 2018, Global road safety status report, 71 p.

UNEP, 2018, The Emissions Gap report 2018, Nairobi, United Nations Environment Programme, 15 p. POCHET Pascal, 2002, "V comme Vélo ou le grand absent des capitales africaines. Xavier Godard", *Le temps de la débrouille et du désordre inventif, Karthala, Inrets* pp. 343-355.

PASSOLI Abelim, DANDONOUGBO Iléri, K. DIZEWE Kossi, AHOLOU Coffi, 2024, "Urban mobility and safety. routière : Approche la sécurité des usagers des modes de déplacement doux dans le Grand Lomé ", In *American Journal of Traffc and Transportation Energineering,* Vol. 9 n°1, pp.9-22.

SUKA Dela Kofi, 2021, *Étalement urbain et problématique de la mobilité dans les métropoles d'Afrique subsaharienne : étude de cas du Grand-Lomé au Togo,* PhD thesis, Department of Geography, University of Lomé, Lomé, 289 p.

Printed by Books on Demand GmbH, Norderstedt / Germany